I0478618

ALTERNATE II

Searching For Answers

Jay Wheeler

Alternate Vol. 2 – Searching For Answers

Authored by Jay Wheeler
Edited by Jay Wheeler
Cover by Jay Wheeler

ISBN-13: 978-1546862888
ISBN-10: 1546862889
1st Edition, 2017

CHAPTERS

INTRODUCTION:

THE MANDELA EFFECT

The first question that needs addressing is "What is the Mandela Effect?". A full breakdown of The Mandela Effect and the possible causes can be found in my first book, *Alternate: The Mandela Effect,* but here is a brief overview...

In the early 1990s, millions of people claimed to have heard and seen reports of Nelson Mandela having died in prison, as well as distinctly remembering news anchors having covered the story, a publicized funeral, and even rioting in the streets shortly after the news broke. However, none of this is true. Apparently.

In 1962, Nelson Rolihlahla Mandela was arrested and sentenced to life in prison. After serving 27 years, Mandela was released in 1990 by President F. W. de Klerk. Many people were shocked because they recall the events previously mentioned.

In 1994, Mandela became president of South Africa. He remained as president of South Africa for 5 years, until 1999. In 2013, he passed away, but when the news broke, more people (those who were unaware of his previous "reincarnation") were shocked because they too claim that they vividly remember his death in the early 1990s.

It was shortly after his actual death in 2013 that this phenomenon really caught steam, maybe due to the internet and people's ability to share their thoughts and memories with the world on a much grander scale than in the 1990s.

Thus, the Mandela Effect was born.

Fiona Broome publicly coined the phrase online in 2010 and she has kept a journal in the form of a website which shares many examples of the Mandela Effect and alternate memories. This phrase has now evolved into a worldwide phenomenon and has captured the attention of millions of people.

Since 2013, there have been many people reporting and sharing other alternate memories that they've had, none of which seem to be unique cases. As soon as a new Mandela Effect example or alternate memory is shared, it seems that many people have the same memory and are baffled as to why they have these memories, despite the fact that they are (supposedly) incorrect. Some of the alternate memory examples reported have a 100% agreeance rate, meaning some of the Mandela Effect examples reported are remembered by everyone in the same way and everyone agrees the example used to be different.

There have been reports of people who have had alternate memories since the 1960s, but it was in 2013, when Mr. Mandela passed away, that things began to snowball. There is even a book titled "*English Alive, 1990: Writings from High Schools in Southern Africa*", in which you will find the following:

"The chaos that erupted in the ranks of the ANC when Nelson Mandela died on the 23rd of July, 1991 brought the January 29th, 1991 Inkatha-ANC peace accord to nothing."

The passage is also in *Western Cape Branch of the South African Council for English Education*, which was published on October 1st, 1991.

If you were under the impression that Nelson Mandela died in prison, you're not alone. However, not everyone has this memory. There are plenty of people that never heard of his supposed death in the early 90s, but there are plenty of examples of the Mandela Effect and I've yet to meet a person that has accurate memories of every example.

Some of the examples we're going to look at are memories that many of us have grown up with and some are popular sayings known the world over. Most will probably make you question the accuracy of your own memory or think of something even more bizarre. Alternate memories are just the beginning. The real meat behind the Mandela Effect is why we're experiencing alternate memories and how it is possible on a global scale. On a quest for answers and closure, this phenomenon and its examples are re-visited, challenged and concluded in *Alternate II: Searching for answers...*

CHAPTER I:

ACCEPTANCE IS UNACCEPTABLE

The reason behind writing this sequel was simply this; acceptance is unacceptable.

Whilst writing the first book, I shared many theories as to why people could be having these alternate memories as well as explaining what the Mandela Effect is, along with a few of the examples. Although there were many pieces to the puzzle and some very detailed explanation, there was no real closure.

Theories range from memory inaccuracy and the brain processing information incorrectly, to simulated reality and the timelines of parallel universes crossing over with ours.

It's great fun learning about the Mandela Effect, testing our memory, recalling our alternate memories, diving into the possibilities of why so many of us have these alternate memories, but that is as far as it goes. Once you have explored deep into the phenomenon, there's still one thing missing - real answers.

In this book, I'll challenge, and more importantly, explain the 25 most well-known and most credible alternate memories people are experiencing.

I will offer complete closure for those who are still searching for answers. I want to point out early, there are going to be completely logical answers for all 25 examples that have become most popular in the Mandela Effect community as well as the reasons for all the other less credible examples that people have scraped together.

If you want to stay in Wonderland and continue to experience the thrill of the chase, you picked up the wrong book.

Everything that you thought might be simulated reality, parallel universes, CERN or "glitches in the Matrix" is about to be answered.

CHAPTER II:

THE HUMAN BRAIN

Before presenting you with answers to the most "famous" Mandela Effect examples, we must ask ourselves how sure we really are of the accuracy of these alternate memories.

Harry Houdini, the world famous magician and escape artist came up with a quote during his reign over the world of magic that went something like this:

"What the eyes see and the ears hear, the mind believes."

Houdini understood that the human brain could be tricked by showing people something other than what was really happening. He achieved this with misdirection. He led his spectators to unknowingly focus their attention on something other than what was actually going on which distracted them from seeing how the illusion was achieved.

Audiences watched as Houdini was shackled upside down in a water tank with the lid fastened with a lock. Within 100 seconds, he freed himself from the locked shackles and escaped the padlocked water tank. And let's not forget his legendary straight jacket escapes. Many consider Houdini a stuntman, but he was more of a magician due to the fact that he amazed and fooled his audience with misdirection and illusion.

The Mandela Effect is not a magic trick or illusion, nor is misdirection purposely used, but Houdini, and many other magicians, are proof that the brain can, and will, whole-heartedly believe something other than the truth until the facts are revealed.

Everyone that watched Houdini's escapes had the same reaction. They accepted what they saw and heard and were adamant that they were correct about what they witnessed. They then saw how the trick was achieved and were completely baffled. Imagine your memories being of what Houdini *wanted* you to see and then being shown what really happened. You were certain you were right until you knew the truth.

Isn't it possible the brain "overlooked" some vital information (the truth) when processing the data that created our memories?

Isn't it possible that our eyes and ears saw and heard something, but the brain processed the data in the easiest possible way and gave us inaccurate memories of that information?

Isn't it possible that after many years, our memories could become faulty?

Isn't it possible that with all of these combined and the influence of other information not true to the source material, the Mandela Effect is nothing more than lyrics, images, phrases and events that we remember just slightly incorrectly?

I just want to open your eyes to the possibility of the inaccuracy of alternate memories before I drag you back down to earth.

I studied magic for many years and I could show you tricks and illusions that would make you question the data your brain processed just moments ago, let alone the recollection of information from years, even decades ago. You see one thing, but something completely different just happened. Just get on YouTube and search for magic tricks. Your brain will be fooled into seeing and believing one thing, while something completely different actually occurred. Of course, the alternate memories were not intentionally designed to trick our brain, but the human brain and its ability to recall past information has proven unreliable before and our brains are wired in pretty much the same way. It would certainly explain why millions of people have the same alternate memories. Most are fooled with simple illusion or magic tricks in exactly the same way.

It's a known fact that the brain processes information in the easiest way possible so it can consume more data. Sometimes it gets things wrong and absorbs data inaccurately.

There are many examples that you can find online that will trick your brain into believing false truths.

One example of this is something called the

"two tables illusion". You can type it into Google and research more information on it, but here is the illustration.

There are two tables shown from different angles that look totally different sizes (You can see the image below). Your brain tells you: "Yep, two tables, one long and narrow and one roughly square".

Have a look for yourself...

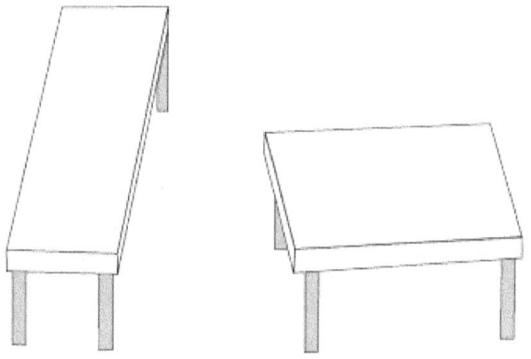

Pretty convincing, right?

Well, these two tables are exactly the same size, both in width and length. If you cut them out and place one on top of the other, you will see their sizes match one another exactly.

Here is an example of both tricking the brain and making it extremely difficult for the brain to work out the most simple equations.

Some people are not fooled or confused by some logical reasoning questions, just as not everyone has false alternate memories, so I will provide a few examples. I have put the answers in the back of the book. Think over the questions and write down your answers before looking in the back. These are extremely simple equations so try and spend no more than 10 seconds on each one. Let's see how you do...

1. How far can a dog run into the woods?
2. A red house has red bricks, a purple house has purple bricks, a green house has?
3. If you dig a hole which is 4 feet wide x 6 feet long and 8 feet deep, how much soil is in the hole?
4. A man is hanging 4 feet from the ground with a noose around his neck. There is nothing but a puddle of water underneath him. He is in a sealed room with nothing else in the room. How did

he hang himself?

5. You fly a plane from New York City to London with 384 passengers. 312 Passengers get off, 294 get on. You fly from London to Milan. 294 passengers get off. What color are the pilot's eyes?
6. Which animal has the ability to reach the top of a melon tree?
7. How many of each animal of each sex did Moses take on his ark?
8. A man lives in a single story home, the walls are made from brick, the floor is made from stone, the ceilings are made from wood. What material are the stairs made of?

Look at the answers to see how many of them you got correct, as well as the ones you didn't, and see how easily the brain can be tricked. If you didn't do well, just remember, it's not a lack of intelligence, it's the brain's ability to process and sort information. Extremely intelligent people recall alternate memories that are incorrect because the human brain is wired in the same way. Here's another example... Spend only 10 seconds reading this phrase and count how many times the letter "F" appears.

FINISHED FILES ARE THE
RESULT OF YEARS OF
SCIENTIFIC STUDY
COMBINED WITH THE
EXPERIENCE OF YEARS.

How many did you count in 10 seconds? 3? 4? The letter "F" actually appears 6 times in this phrase. The brain is designed so you can read the information quick enough, but it commonly overlooks information, even when you know what you are looking for.

By now, you're hopefully beginning to see how unreliable the brain is when processing data. It can be wrong, it can be fooled, and it can certainly overlook the most basic information. These examples will make the next chapter easier to stomach if you're a die hard matrix or parallel universe theory supporter.

Understanding that the brain is not perfect,

as *nearly* everyone gets tricked by the table illusion and other examples that were given, I hope I have made it easier to believe that *nearly* everyone has the same alternate memories of events that really didn't go down the way we *believe* we remember.

The human brain has been proven inaccurate many times over and surely will be many more times in the future. Our shared memories are simply a product of inaccuracies in the data collected by our brains, fast or poor processing of information by the brain and influence from other things and/or society.

Now that you're warmed up, let's get the answers to this mind-bending phenomenon.

CHAPTER III:

CHALLENGING ALTERNATE MEMORIES

In this chapter, we will jump onto the side of the coin that many people are not inclined to believe. Many of us want to believe there is something spectacular going on that can not be explained.

Forrest Fenn, a wealthy art collector from New Mexico, allegedly hid millions of dollars worth of gold, silver and precious stones in a chest somewhere in the Rocky Mountains with clues leading to its whereabouts. It is still out there, according to him.

He wrote and titled a book: *"The Thrill of the Chase"*.

Searching for answers and chasing the unknown is a thrill, but for some, closure and answers are more important. That's why I decided to drill down into the most talked about Mandela Effect examples and provide all the answers you have been looking for.

Have you ever been 100% certain about something in life and then you find out you were completely wrong?

Is it really that difficult to believe that the big cool name of THE MANDELA EFFECT has overshadowed the absolute simplicity and possibility of a letter or word that you may remember incorrectly from years ago?

Think about it. We forget where we put our keys 10 minutes ago. We forget to write things down. We forget things daily.

Have you ever said, *"I've done it again"* when you realize you've made a mistake that you've made before? Have you ever had to go back and check something you saw seconds ago because you've already forgotten? Have you ever thought or spoken about something for a long duration then said to someone *"refresh my memory"* or *"remind me again"*?

Strip away the cool name this phenomenon has been given, strip away the hype, strip away the community of people who share these stories and all you are left with is our brains' common ability to make very simple (and relatively unimportant) mistakes over the years as it has been collecting millions of gigabytes of data and information.

The brain is an incredible tool, but it is far from flawless. It is important to understand this because people who want to believe the Mandela Effect is real, may have a tough time getting through this chapter and realizing so much of their memory is incorrect.

The brain is not a computer with a hard-drive that can remember everything perfectly. There's limited storage and it will be selective as to what information it deems important.

If you took the ten most intelligent people on planet Earth and gave them the top 25 Mandela Effect examples, as well as the brain teaser mentioned earlier, it's extremely unlikely a single one of them would have 100% accurate memories of those examples. I say this because what I'm sharing is not to suggest that anyone is stupid, it's just simple imperfections in the brain's ability to remember that we all share.

I am going to eliminate all reported alternate memories that are small and insignificant, because those will become trivial by the end of this chapter, such as the KitKat logo without the hyphen or the "s" missing from the "Depend" brand. We're going to go over the alternate memories that really have some significance to them, rather than the more easily mistaken examples that some may consider scraping the barrel for hype.

You're about to get a more down to earth explanation for all the alternate memories that have baffled the world.

Some may believe something far more advanced than our civilization is pulling the strings. Either way, I hope you find some of the answers you are looking for...

#1 – Nelson Mandela

The number one alternate memory to tackle has to go to the date of Nelson Mandela's death. Many people remember that he died during his time spent in prison in the early 90's. There is even a book titled **English Alive, 1990: Writings from High Schools in Southern Africa**, in which you will find the following:

"The chaos that erupted in the ranks of the ANC when Nelson Mandela died on the 23rd of July, 1991 brought the January 29th, 1991 Inkatha-ANC peace accord to nothing."

The passage is also in **Western Cape Branch of the South African Council for English Education**, which was published on October 1st, 1991.

It is important to remember there was no mainstream internet during this time, so a rumor of Nelson Mandela's supposed death could have spread like wild-fire via word-of-mouth with no real ability to check or correct the accuracy of the rumor. Having a few people misinformed about his death, then publishing it in books, it is easy to see how so many people have built memories of this event that are not true.

If you're not a follower of South African ruling, you may not have ever come across the truth of his release from prison or his promotion to president of South Africa, which is why the global news of his death in 2013 was such a shock to many people who claim to have memories of his death over 20 years prior to his actual death.

#2 – Empire Strikes Back

The Darth Vader quote to Luke from *Star Wars: Episode V – The Empire Strikes Back* has been a very hot topic for Mandela Effect researchers and Star Wars fans alike.

Many people remember Vader's line as, *"Luke, I am your father"*, but if you watch any version of the movie now, the line is, *"No, I am your father"*. Many are adamant that this line has somehow changed, but if we examine a bit more script from the motion picture, this is how the scene goes:

Vader: *Obi-Wan never told you what happened to your father...*
Luke: *He told me enough, he told me you killed him.*
Vader: *No, I am your father.*

It is easy to argue that, *"No, I am your father"*, is a more natural reply than, *"Luke, I am your father"*, because Vader replies to Luke's comment and corrects him by saying "no" (I did not kill your father), then reveals the truth.

There is a point, later in the movie, where Vader just says, *"Luke"*, when using the force to connect with him, and Luke replies, *"Father"*.

It is possible that we have confused these two parts of the movie, making us think the line was, *"Luke, I am your father"*.

We could also look at the fact that the scene was such a powerful moment in the movie that the attention was taken off whether he said *"no"* or *"Luke"* and the attention was focused on the big reveal that Vader was Luke's father.

Also, before the days of the internet, having a few people incorrectly spread the quote as *"Luke, I am your father"* could have caused many to not have realized that the quote was slightly inaccurate. Without an online community of people to agree on the line, it's possible millions of us just accepted the line as we all "remember" it. The false memory has been so deeply embedded into our culture, that it's extremely convincing that the line in the movie script, Blu-ray, DVD, and original VHS has magically changed.

Lastly, the original storybook from the movie that was released around the same time has Vader's line as, *"No, Luke, I am your father"*. There is an image from the book which you can see online: **https://goo.gl/rZnYKe**

#3 – Three Little Pigs

There have been many online debates about the classic nursery rhyme of the *Three Little Pigs*, and whether the line is, *"I'll huff and I'll puff and I'll blow your house down"* or *"I'll huff and I'll puff and I'll blow your house in"*.

This "alternate memory" is simple to explain. I am from the UK, where I was raised and where I heard the story of the *Three Little Pigs* continuously throughout my childhood. I know the line as, *"I'll huff and I'll puff and I'll blow your house down"*. I not only had the books read to me, but I read the books myself. But many people I have spoken to in America know the line as, *"I'll huff and I'll puff and I'll blow your house in"*.

In different geographical locations, the dialect and terminology can be vastly different. Depending on where you live/are from, there are two versions of the line. This alternate memory is very split between people in different locations around the world and no one can agree on one because both have been widely used for many years.

There is a British version and an American version, as well as versions from other countries. There is a *Three Little Pigs* rhyme and a *Three Little Pigs* story. The confusion has come from both the story and the rhyme, as well as versions from multiple geographical locations. Many people on the internet just fail to ask where the other people debating the alternate memory have grown up and acquired the memory from.

#4 – Chic-fil-A

Considered one of the greatest Mandela Effects to date (at least for Americans) is the Chic-fil-A logo. This has been the backbone of nearly every conversation I have had about the Mandela Effect. So, let's get into it...

For years, I saw the logo and always thought it was "Chic-Fil-A". Then, when I first heard of the Mandela Effect, I was thinking it was all a load of people bending the truth to get views on YouTube or traffic to their articles, until I came across the "Chic/k-fil-A" battle.

Needless to say, I was blown away. I immediately researched online to see if there really was a "k" in the company name. I then contacted corporate to ask them when they had added the "k" in, so I could put this Mandela Effect rubbish to rest. If you read my first book, you'll know the answer I got did not putting anything to rest. Customer service told me that their company name has always had the "k" in it and has always been spelled "Chick-fil-A".

If you're American, or live in the USA, there's a strong chance you have been surprised by this example, too. Here's the conclusion to this highly debated Mandela Effect example...

Most people remember it as "Chic-fil-A", but there are some who remember it as "Chik-fil-A" and a few as its currently spelled (Chick-fil-A).

Another possibility may be that the font on their logo is a playful one with a calligraphy style and could serve as an eye illusion to many people, which would make us process it without the "c" or the "k".

If the spelling has always been "Chick-fil-A", our brains have processed "Chick-fil-A" as "Chic-fil-A" because of the logo's font and because our eyes have deceived us. It's possible the font may have also disguised the "c", making people think it was "Chik-fil-A" and for some, the font disguised the "k" and we processed it as "Chic-fil-A".

The company name was registered in 1963 with the first location opening in 1967, during which time, the name was registered as, and has always been, "Chick-fil-A". Due to the common misspelling of the name, this is why the company purchased "chic-fil-a.com" and re-directed it to their website, chick-fil-a.com. The official company website, chick-fil-a.com, was purchased on November 25th, 1995, while chic-fil-a.com was purchased on November 4th, 2002. This was around the same time the internet became mainstream and was standard in households across America.

Google Trends will not allow you to see word searches from 2002, but I am confident in guessing that there were so many searches for "Chic-fil-A" and other misspelled variations of the company name when the internet became mainstream, that Chick-fil-A decided to acquire misspelled website domains and link them to their official home page.

Furthermore, some of the company's packaging over the years has had the logo and company name printed on the front, but due to inaccurate cropping, the "Chic" was printed on the front, while the "k" was on the side and hidden from the main area the eyes would focus on. This would build a strong belief in the name being "Chic" when paired with the font that may have accidentally tricked our brains into overlooking a letter, just like the table illusion and other brain tricks.

#5 – Pikachu's Tail

Pikachu's tail has been discussed as a Mandela Effect example. Some believe his tail used to have a black tip, which is now gone. People are very divided on this example and at least half of people do not share this alternate memory, many of which are die-hard Pokemon fans. My simple answer is, Pikachu's tail has always been solid yellow with no black tip and people are confusing the black tips on his ears with a black tip on the tail.

Of course, we see what we want to see, so it's entirely possible that those who remember a black tip on the tail were victims of an accidental eye illusion or the brain processing information slightly incorrectly.

#6 – Looney Tunes

Another easily explained Mandela Effect example is the Looney Tunes logo. Many believe it used to be spelled "Looney Toons", but the people who remember this are likely getting it confused with "Tiny Toons". Tiny Toons is spelled with the "oo" while Looney Tunes is spelled with "u". Easy mistake.

#7 – The Berenstain Bears

Another example that has a lot of attention on it is *The Berenstain Bears*. Many people remember the classic series as being called "The Berenstein Bears" with an "e" instead of an "a".

With the deceptive nature of the brain now being something we can consider, as well as the information shared regarding the "Chick-fil-A" logo font, it is entirely plausible that *The Berenstain Bears* has always been spelled with the "a" and our brains perceived it as "Berenstein" for ease of processing, as not everyone remembers it as "Berenstein" and have no alternate memory of this example. A family member of the creators even came out and said, *"Of course it has always been Berenstain"*. It is strange to think one of the most talked about Mandela Effect examples has stemmed from a single vowel recalled from years ago.

#8 – The Silence of the Lambs

One of the incredibly confusing Mandela Effects is a line from *The Silence of the Lambs* - or lack of it! Many remember a line from the movie where Hannibal Lecter says, *"Hello, Clarice"*, but if you watch the movie again now, the famous line, that we all know, is not there and never has been. So, how is it possible that we all remember it?

Just because the line was never in *The Silence of the Lambs*, doesn't mean you've never heard the line. You probably have, so you're not remembering something that never happened, you're remembering something that did happen, just not in the film *The Silence of the Lambs*.

If you watch the 1996 movie *The Cable Guy,* with Jim Carrey, there is a scene where he and Steven (played by Matthew Broderick) visit a themed restaurant called Medieval Times. During the scene, Jim Carrey asks for Steven's chicken skin and places it on his face. He then says, *"The Silence... of the Lambs."* - *"Hello, Clarice. It's good to see you again."*

Jim Carrey is a master of impressions and he says the line in Hannibal's voice. *The Cable Guy* came out just a few years after the initial release of *The Silence of the Lambs* and it was almost as big of a hit. *The Cable Guy* is not the only movie/show that utilizes the line, *"Hello, Clarice"*. It has been said multiple times and there have been many parodies of it over the years.

With years past and the line so deeply

embedded into our culture, it's very easy to see why it has become such a talked about Mandela Effect example.

Don't worry, I, too created an image in my head of Hannibal saying the line, but it was a fabrication of the truth. The line was said in a movie and it was said in Hannibal's voice, just not in the movie we thought.

#9 – JCPenney

The department store, JCPenney, has become a Mandela Effect example as many people remember it as "JCPenny", without the additional "e" in the name. It's not difficult to imagine this additional "e" was overlooked in the logo.

Everyone has seen the word "penny" a million times throughout the years, when you add that fact to easily overlooking the spelling of the company name, JCPenney, it might seem as if you have an alternate memory.

Although a red flag was raised for many when the actual spelling of the company was brought to people's attention, chances are, you could have continued to see the logo and company name for the rest of your life without ever giving it a second glance. Unless you're looking for it, you probably won't notice it if it's not significantly out of the ordinary.

#10 – Cadbury's Creme Egg

As with the *JCPenney* example above, we have all seen, heard and spelled the word "cream" numerous times in our lives. Again, without someone having pointed out the spelling of the word "cream" in Cadbury's Creme Eggs as "Creme", you may never even have noticed it. The spelling could well be something the brain overlooked and did not *notice* the word was spelled differently until there was attention drawn to it by others who picked up on it.

#11 – Largest U.S State

There has been a lot of debate about which U.S state is the largest. I, for one, had to do some research into this, not only to find out the largest U.S state, but why this is even considered a Mandela Effect example or an alternate memory.

When I was very young and in school (in the UK), I was under the impression that Texas was the largest U.S state. A few years later, as a young teenager, I remember playing *Trivial Pursuit* and the question came up: *"What is the largest state in America?"*. Obviously, I answered Texas, but the answer was California.

I recently realized that the largest state in America is actually Alaska (and the largest by quite a bit). Others have also run into alternate memories when answering the question. Here's why...

I, and anyone else who thought Texas was the biggest state, was clearly just misinformed. California is the largest U.S state by population and Alaska is the biggest state by land mass. It appears the inaccuracy came from the question and the confusion between biggest land mass and largest population. This was likely the reason for it having been added as a Mandela Effect example.

#12 – Beats By Dre

The popular headphones, Beats by Dr. Dre, have been commonly referred to as "Beats by Dre", as it's faster to say and everyone knows Beats by Dre were created by the rapper Dr. Dre.

It seems the confusion has come from the logo "Beats by Dr. Dre" and people most commonly using the shorter term "Beats by Dre". It's possible that people have become so used to the shortened term that when reading the full name on the product, it appears out of place, but many will agree the name has not changed from "Beats by Dre", rather, it has always been "Beats by Dr. Dre" and just more simply referred to as "Beats by Dre".

The official website is beatsbydre.com which may have caused further confusion. The website beatsbydrdre.com re-directs to the main website and that may have raised even more suspicion, but that is, more than likely, just for SEO (search engine optimization) because people would have searched "beats by dr dre" or "dr dre beats" when they were first released and not widely known as Beats by Dre. And if you think about it, why wouldn't they own both domain names?

#13 – The Matrix

What if I told you "*What if I told you...*" was never in The Matrix?

Millions of people remember Morpheus delivering a line to Neo during their first meeting that went something like, *"What if I told you..."*. Some say it was: *"What if I told you, everything you knew was a lie"*, others remember it as: *"What if I told you, everything you know to be true is wrong"*. The truth is, neither line was ever in the movie, nor was the famous *"What if I told you..."* in any context. This Mandela Effect example is the product of the internet and nothing more. For years now, there have been memes of Morpheus that start with *"What if I told you..."*. It's a line that seems very fitting for the movie, but it was created for a meme and never featured in the movie. If you Google search "what if i told you meme", you will see that there are two main pictures that are the background for the meme. Both are of Morpheus, one of which is while he is sat in the red chair during he and Neo's first meeting, the other is of Morpheus while he is inside the Matrix with Neo. Of course, there are multiple backgrounds used because the line, *"What if I told you..."*, is not linked to either of those scenes.

Morpheus in the red chair, during he and Neo's first meeting is the most commonly used image for the memes and that's probably why people expect to hear that line during the scene where he first discusses the Matrix with Neo.

The first person to create a meme with that line was in 2001, a couple of years after the movie's release. The fact that the movie came out so long ago and your brain not connecting that specific line when re-watching the movie (if you have re-watched the movie in the last 18 years) would not necessarily be noticed, but when someone tells you that the line you know so well is not actually in the film and you look for it, that hardwired quote in our brain becomes a Mandela Effect for a majority of people who have been influenced by the internet memes instead of the actual source material.

#14 – Chuck E. Cheese's

This Mandela Effect is just like the "Beats by Dre" example. It's easy enough to believe that many people would simply say "Chuck E. Cheese" as opposed to "Chuck E. Cheese's". The company claims their name has always been Chuck E. Cheese's, but due to the fact that "Chuck E. Cheese" is just a little easier to say, it's become embedded in our memory banks and we pay no attention to the name when we see it until we are made aware that there is an "s" on the end.

#15 – Dolly's Braces

In the 1979 movie, *Moonraker*, many people remember Dolly (Jaws' love interest) with braces on her teeth, but upon re-watching the movie, it is clear she does not have braces.

The first report of this inaccurate / alternate memory was from August 2nd, 1999 which you can see here: **https://goo.gl/TEvXy5**
Over the last couple of years, the example has received a lot of attention within the Mandela Effect community, but there is very little in the way of answers to why many people are experiencing this memory when it's clear that Dolly does not have braces in the movie.

From the 1999 post, we can see that this is not a new example as people have been claiming to have alternate memories of Dolly with braces for some time. There are a few reasons that could provide answers as to why people might think they remember it that way:

> 1. Jaws and the love affair that spawned, seemingly from her first smile. People

seem to remember the love affair started as soon as she smiled and Jaws saw her braces. Back in 1979, the VHS quality was poor in comparison to DVD, let alone Blu-ray. The first release of *Moonraker* was probably in a resolution no greater than (or equivalent to) around 450x450 pixels. With poor lighting and the shine on her teeth, some people may have thought she had braces.

2. In the scene where she is drinking champagne with Jaws, it looks like she does have braces through the reflections in the glass which could have added to the belief.

3. Another reason is, well, braces would completely suit her and it seems natural she would have them. A young girl in the 70's with freckles, pigtails and glasses, you would almost assume she had braces, even if she never opened her mouth. Combine that with the fact that Jaws had metal teeth and the brain will logically arrive at

remembering her with braces.

4. There have been many fan sites and websites that have profiled the character over the years, all of which incorrectly describe her with braces. This adds to the belief that she had them in the movie.

5. Finally, in 1991, sometime before the first alternate memory was reported, there was a Visa commercial that featured Richard Kiel (the actor who played Jaws). The Visa commercial promoted the mini-card and the cashier in the commercial had a very similar look to Dolly's character in *Moonraker.* She was blonde with an innocent look and the camera angle of the cashier was nearly identical to the camera angle of Dolly when she first met Jaws in *Moonraker.* The cashier looked up at Richard Kiel and smiled at him. Her smile revealed she had braces which was clearly a little "wink" at Richard Kiel's character of Jaws over a decade earlier. If you do a Google search, you

can easily find both stills from the movie and the commercial. It's easy to see where the confusion came from.

If you put all these pieces together, you won't need to stretch the imagination much to arrive at the conclusion that people who remember Dolly with braces may have been mistaken and are recalling false memories from decades ago.

#16 – Interview with *the* Vampire

Another largely discussed alternate memory is *Interview with the Vampire*. This is likely just a mis-remembered word as it's just "*the*" changed from "*a*". Although it's a very small difference, it has caught a lot of attention.

The German movie title is *"Interview mit einem Vampir"* which directly translates to *"Interview with a Vampire"*. The French title also translates directly to *"Interview with a Vampire"*.

It's not difficult to see the confusion could

have come from the French and German versions and become our belief of the movie's title over the years. It's a very minor difference which could have been easily overlooked until it was brought to our attention.

Even Amazon employees have got "*a*" and "*the*" mixed up as you can see from the listings of the movie on their site. Here's a screen shot of the search results for "*Interview with a Vampire*" on their website from July 2017: https://goo.gl/iVTDDB

#17 – Monopoly Man's Monocle

Another of the most talked about alternate memories is Rich Uncle Pennybags (the Monopoly guy) or, more accurately, his monocle. Nearly everyone who has discussed the subject remembers him as having a monocle (eye glass), but every image of the famous board game character shows him without a monocle and he's apparently never had one. So why do so many people seem to have the memory of Pennybags with a monocle?

Do you really remember the Monopoly man with a monocle or have you seen *Ace Ventura: Pet Detective*? Another movie starring Jim Carrey that takes and twists source material and makes it its own. When an older, bald man walks down the stairs wearing a black and white three piece suit, Ace refers to him as "The Monopoly Guy". In the movie, the man wears a monocle. This reference is what is probably responsible for millions of people believing that the Monopoly man is meant to have a monocle.

There are many other similar characters that all wear monocles, including the Penguin from Batman, Mr. Peanut, and the Mayor from The Power Puff Girls.

Similar to Dolly from Jaws, the Monopoly man seems to be the kind of character who would suit a monocle. He's an old fashioned character with a three-piece suit, a top hat and a cane. It's logical to think a monocle would be the perfect piece to complete his outfit. There is even the odd piece of merchandise where he does have a monocle, if you search hard enough.

Monopoly actually discovered the confusion and posted a picture of the Monopoly man with a monocle in May 2016 which you can see here: **goo.gl/gC1yt3**

#18 – Sex *and* the City

The *Sex and the City* alternate memory is simply due to the different names between the TV show and the perfume. The TV show is called *Sex **and** the City* while the perfume is called *Sex **in** the City*.

#19 – Alice in Wonderland
"We're all mad here"

The Cheshire Cat from the animated Disney classic, *Alice in Wonderland*, is remembered by a large number of people to have said the quote, *"We're all mad here"*, but as it turns out, the line is actually, *"Most everyone's mad here"*. There are many people who remember the quote so vividly that they even had the quote: *"We're all mad here"* tattooed on themselves.

So, has there been some sort of glitch in the Matrix or a timeline interference from a parallel universe? Probably not.

This is just another inaccurate memory.

Firstly, if you conduct a Google search for **"we're all mad here quote"**, you will see there are two slight variations, one being, *"We're all mad here",* and the other being, *"We are all mad here"*.

The important thing to note here is that Disney's 1951 animated *Alice in Wonderland* movie is only one version and is not where the story originated. Lewis Carroll first wrote the book titled *Alice's Adventures in Wonderland* in 1865. You can see it here: http://amzn.to/2sMvZ9i

There is at least one more version of the *Alice in Wonderland* story in the form of a book by Jane Carruth, where the Cheshire Cat says the line, *"We're all mad here"*. You can see it here: http://amzn.to/2spAasp

Understandably, there have been literally thousands of parodies, plays and reenactments of the *Alice in Wonderland* story, many of which use the line from the original story. This would have influenced millions into believing that the line had been in Disney's 1951 version when it never was.

As for people whom have had the quote tattooed on them... The tattoos are written as *"We're all mad here"* because it's just text, but no one has a tattoo of Pikachu with a black tip on his tail because the tattoo is copied from a picture, where as the quotes are written out by the person getting the tattoo and not copied from the video footage from the movie or the script. Many people probably just did a Google search to get the quote and got it tattooed onto their bodies, believing that it was the exact quote from the Disney movie.

The same thing would happen if someone did a Google search to find Morpheus' supposed line, *"What if I told you"*. They would find a million memes with the line, but it would not be from the actual film.

The Cheshire Cat's famous quote *is* from the *Alice in Wonderland* story, just not the Disney version.

#20 – I'm a Barbie Girl

In the well-known *Aqua* song "Barbie Girl", there is one simple little lyric that has been questioned by the Mandela Effect community. Many people remember the line as, *"I'm a Barbie girl, in **a** Barbie world"*, but the line is actually *"I'm a Barbie girl, in **the** Barbie world"*.

It's yet another very subtle difference between the words "*the*" and "*a*". It is logical to think that the line was simply *misheard* as "***in a***" sounds very similar to "***in the***" when being sung in a song, due to the way that lyrics often run together to make everything flow just right.

If you go back and listen to the song again, you will hear the line multiple times throughout the song and the way it is said can very easily be mistaken as, *"In **a** Barbie*

World", rather than, *"In **the** Barbie world"*. The music playing, the tempo, and the female singer's accent really does disguise the word "*the*" and it can easily be heard as "*a*", which is why it has been (mistakenly) added to the Mandela Effect phenomenon.

#21 – Jaws – Bigger Boat

Regarding the 1975 Spielberg hit, *Jaws*, there has been some debate as to whether Chief Brody (played by the late Roy Scheider) says, *"**We're** gonna need a bigger boat"* or *"**You're** gonna need a bigger boat"*.

A good portion of the Mandela Effect community remember it as *"**We're** gonna need a bigger boat"*, which is logical considering Brody is on the boat and is part of the team hunting the enormous, man-eating shark. It's just as easy, however, to understand that he might say, *"**You're** gonna need a bigger boat"*, as he was speaking directly to the owner of the boat after having seen the shark emerge from the water for the first time. Both lines actually make sense, but the line has always been: *"**You're** gonna need a bigger boat"*.

Once again, the internet plays a large part in this one. The movie is over 40 years old and the sound quality of movies from decades ago was nowhere near as good as it is today. The character of Brody also had a cigarette in his mouth as he delivered the famous line which likely also contributed to people having heard it wrong.

People having misquoted the line for years and years has caused it to become embedded in film culture, so when you listen more closely and you hear the actual line, it appears to be incorrect when that is just simply not the case and there is no evidence that it ever was.

#22 – Beam me up, Scotty

Despite the fact that the famous quote, *"Beam me up, Scotty"*, tends to be tightly connected to *Star Trek*, the line was never actually in the television series. It's simply a misquoted catchphrase that has made its way into pop culture.

William Shatner has quoted similar lines, but the one everyone knows so well, *"Beam me up, Scotty"*, is one that was never used in the television series. Those who are adamant it was in the original show have created false memories and visualizations of the character having said the line, when in reality, the influence came from somewhere other than the direct source material, just as with the quotes from *The Matrix* and *The Silence of the Lambs*.

#23 – The Lion and the Lamb

When I discuss and share the Mandela Effect with people, I always arrive at this example, which never fails to astonish people. Nearly everyone I've spoken with who knows anything about the Bible remembers the lion as having been the animal that is said to lay down with the lamb. I think I have only spoken to one person that actually remembers this example correctly, everyone else has an alternate memory.

It was actually never a lion, it was always a wolf.

Just like with the movie quote examples and a loose connection to the source material, the truth has been bent over the years by storytellers, pictures and illustrations and the incorrect information has been ingrained in the minds of millions of people.

The lamb laid with a wolf in the Bible. If you read the Bible, it's there in black and white. If you can find someone who has read the Bible many times without outside interference or cultural influence, they will also be able to tell you it was a wolf that the lamb was said to lay down with and not a lion.

#24 – Eli Whitney

Eli Whitney was an American inventor who lived from 1765-1825. He was a white man.

George Washington Carver was an American inventor who lived from the 1860's-1943. He was a black man.

There has been a little confusion over Eli Whitney's ethnicity in the Mandela Effect community, but it appears that people may be confusing him with George Washington Carver.

Although they were alive 40 years apart, they could still be considered to have been from the same era.

Once someone shared pictures online of Eli Whitney, there was a hoard of people that took to Twitter, clearly stating they had been taught in schools across the nations that he had been a black man. One person even claimed that they had passed a history test and got the answer correct by stating he was a black man. This is not true and never has been. This can be chalked up to simple confusion as both men were inventors (and maybe poor education for the unfortunate person who supposedly got an incorrect answer "right" on their history test).

#25 – Snow White and the Seven Dwarfs

This example is one of the cornerstones of the Mandela Effect and it's one of the best, which is why it was saved until last. It may be the Mandela Effect example and alternate memory that has received the most attention. It is one of the most baffling, but like the other previously unexplained alternate memories, it's time to bring the fairytale to an end.

In the original 1938 Disney classic, *Snow White and the Seven Dwarfs*, the Evil Queen walks up to her magic mirror and asks it a question. It's just that simple.

The line goes like this: *"**Magic mirror** on the wall, who is the fairest one of all?"*.

So why does nearly everyone on planet Earth remember it as: *"**Mirror, mirror,** on the wall..."*?

We must remember, this animated version of *Snow White and the Seven Dwarfs* was released in 1938, getting close to a century ago. When was the last time you actually watched Disney's

Snow White and the Seven Dwarfs prior to hearing about the Mandela Effect? It's been quite a long time for most of us, I'm sure.

I think most of us have heard someone repeat the quote since then, whether it be in other movies or shows or by other individuals. Once again, we've most likely been influenced by information that is not accurate and over the years, it has become so deeply hardwired into our brains that even when we have gone back and re-watched the film, most of us would have easily glazed right over what the Evil Queen actually said.

So, if the line is indeed *"Magic mirror"* and not *"Mirror, mirror"*, where did the *"Mirror, mirror"* phrase even come from?

One logical answer to this question could be that the original Brother's Grimm story, *Little Snow-White* by Jacob and Wilhelm Grimm, tells the story of a king's beautiful wife, the queen, who stood before her magic mirror every morning and asked:
*"**Mirror, mirror**, on the wall, who in this land is fairest of all?"*

It may be difficult to believe we've been wrong about the famous line having been in the animated Disney classic we all know and love, but there is enough evidence to logically conclude that it is simply a case of mis-remembering and misinformation. Of course, it only adds to the confusion that there are a vast number of products on the market bearing the misquoted line.

So the "*Mirror, mirror*" phrase does appear in the story of *Snow White*, just not in Disney's animated version like we all thought we remembered.

Conclusion

Now that we've broken down the top 25 Mandela Effect examples, it can be difficult to come to copes with some of these answers when you feel so sure of what you believe you remember, but if you will allow yourself to think logically, you will realize that the Mandela Effect is just a hype-word for something that has been used by conspiracy theorists to get everyone bent out of shape over a "phenomenon" that can be attributed to simple human error and inaccurate memory.

Chapter IV: Understanding

The previous chapter may have been difficult to accept for some people. I truly hope you have the answers you were looking for and that you aren't too disappointed that the Mandela Effect isn't the mind-bending phenomenon that it's been made out to be.

The brain, in many ways, is just like a computer. When you overload a computer with information, be it opening too many programs, too many browser windows or downloading too much data at once, it will shut down and potentially lose some of the data it had already acquired. The brain does something similar. It will retain important information, but overload it with data, like most of our brains are nowadays, and it will lose bits of information along the way. When you attempt to recall memories from the past, especially if it is something from the distant past, you may remember some things inaccurately or have pieces of information missing from these memories.

It is said that the average person will see 2,000 ads per day. That's 14,000 a week. Not to mention, the several emotions, some 70,000 thoughts, and all the other data our brains consume in a single day.

The brain has between 80-100 billion neurons, each of which fire multiple times per second. That means your neurons are firing 34.5 quadrillion times per day. Let's put that into some perspective.

- 34.5 quadrillion dollars = 388,000 times richer than Bill Gates.
- 34.5 quadrillion miles = A trip to the moon and back, 72 billion times.

Scientists say that each time we remember something, we re-construct the event from trace throughout the brain. Our memories could also be adaptive, re-shaping memories to accommodate new information and situations.

Ultimately, we could say that our memory is somewhat *flexible*.

The truth is, we experience our own individual alternate memories all the time, but we brush them off as simple mistakes and move on.

Have you ever thought you locked the door before bed, but you actually hadn't? Maybe you have been so sure that you put your keys in a certain place, but then discovered them elsewhere? Have you ever walked into a room and completely forgotten what you went in there for? Or maybe you were having a conversation and you completely lost your train of thought and totally forgot what you had been talking about only moments before.

These are just a few examples of memory inaccuracies that we experience all the time, proof that we can forget things in literally seconds!

To take it a little further, you may have used a recipe or repeated something hundreds, or even thousands, of times in the past. When some time passes, you try to do that same activity again and you've completely forgotten how to do it. People experience this everyday. So, if this is possible and it's not uncommon, why is it so difficult to believe that our memories of a movie we've seen a few times, years ago, might be slightly inaccurate?

Well, it's not, but the cool name and the hype has built it up to be something out of the ordinary. This, combined with many people's lack of knowledge in psychology, neuroscience and the functionality of the brain has led to the false belief that there is some kind of "glitch in the Matrix" that some believe we are all living in that is causing the phenomenon known as the Mandela Effect.

Having said all of this and the fact that the Mandela Effect is not *alternate* memories, but *inaccurate* memories, doesn't necessarily mean everything is as it seems.

Chapter IV:
Beyond
Understanding

In a conversation a million miles away from the Mandela Effect, I believe there is far more to our reality than we understand.

Do I believe the Mandela Effect is a glitch in the Matrix? Do I believe the Mandela Effect is the timeline of a parallel universe crossing over with ours? Do I believe CERN's Large Hadron Collider or something involving experimentation with quantum mechanics is responsible for the Mandela Effect? Absolutely not. And hopefully by now, you don't either. But the Mandela Effect is not the only evidence that there may be something more to our reality, or even, beyond our reality than what we know.

This chapter is mainly for people who were disappointed to find out that the Mandela

Effect isn't what we've been led to believe. There are still plenty more questions that have yet to be answered about our reality, consciousness, the universe, and life after death, so I am going to share some quick-fire thoughts on a few subjects that may reinvigorate your curiosity and interest in life's (still) greatest unsolved mysteries and unanswered questions.

Life After Death

Even with all of the grand technology that we have today, we still don't know if there is life after death. Some have faith in a higher power, a Creator, and believe that there is a heaven and a hell. On the opposite end of the spectrum, there are those that believe we lose consciousness when we die, our brain completely shuts down and we simply don't exist in any form or fashion from there-on. Then there are others who believe our energy (or soul) is re-incarnated into another body while everything previously stored in our physical brain, such as memories, is wiped clean and we begin a new life. Whatever the case, I believe, due to our unique ability to reason, the fact that we are self-aware, and the way we technologically evolve when other earth-dwelling creatures do not, it is more likely that death is not the end for us.

Whether we pass into a higher dimension that we can not (yet) access while we physically exist in this world or whether we go to heaven or hell, there are a few more reasons to believe that death is not really the end.

Firstly, if there is no life after death, every single paranormal experience, report and case in human history must either all be hoaxes or a massive number of people throughout history must be and have been vastly mistaken and/or incredibly gullible.

Sure, there are hoaxes and times when people elaborate on the truth for show, but I find it unlikely that every single paranormal event ever reported is explainable within the confines of our reality and our understanding of it.

The paranormal does allow for the

possibility of other dimensions (or entities thereof) having the ability to interfere with ours, however, I don't believe that that has anything to do with the Mandela Effect, as most examples that have been given can be explained relatively easily.

Life Beyond Earth

Probably the number one subject in the unexplained arena is that of aliens and possibly the biggest question we face, behind "what is the meaning of life?", is "do extraterrestrials exist?".

This warrants an entire book on its own because I have done extensive research into this subject, but in a nutshell, I, myself, do believe in intelligent life beyond earth.

A largely discussed theory of the Mandela Effect is that we are living in a simulation. A simulation which would have to be controlled or run by some kind of advanced civilization.

Due to poor imagery, movies, and other sources, aliens are expected to be like *E.T*, the aliens from *Mars Attacks!,* or the *Alien* franchise. This makes many people laugh at the idea of "aliens" living somewhere in the universe and even more so at the idea that they've been visiting our planet. Some people even claim to have been abducted by extra-terrestrials. I feel these things have led people to consider the subject laughable, although, lately, the idea seems to be becoming more widely accepted and talked about in a serious manner.

If we filter out Hollywood's visualizations of aliens and other portrayals, the conversation can be very serious when discussed in a scientific manner.

Let's start with the law of probability. It is probable and logical, that in a universe with billions and billions of light years of seemingly untouched real estate out there, that there would be other intelligent life. And of course, we haven't found the edge of the universe and the possibility for multiple universes is being seriously considered by the scientific community.

Secondly, we can look at something similar to the paranormal category...

Every single unexplained light in the sky, the Phoenix lights, the 1947 Roswell incident, *The Interrupted Journey* and many more reports would all have to be complete hoaxes or completely explainable, which hasn't been the case.

A more recent event that really caught my attention was in September 2016, when Elon Musk's rocket strangely exploded on the launch pad. The story that came out of SpaceX was that it involved liquid helium, advanced carbon fiber composites, and oxygen so cold that it actually enters solid phase.

It took them some time to figure out the cause for the explosion. Before they released the apparent reason for the explosion, there was a rumor going around about static energy.

Anyway, the reason it caught my attention is because, after having played back and watched the video of the explosion numerous times over, I could see a small black object flying toward the rocket at extremely high speeds (much faster than any man-made vehicle) at the exact moment of the explosion. The black object then changed direction after the explosion. The object flies from the right and goes behind the two structures. Judging by its distance from the camera, you can pretty well calculate that the object is traveling far too quickly to be any man-made craft or bird that most of us know about. The size is a lot smaller than a plane as well. It appears to be big enough that it could transport a couple of humans within it. You can see the video here: https://goo.gl/D3e2Bo

If you click the settings icon on the YouTube video, you can click "speed" and then choose "0.25" to slow the video down and decide for yourself. You can use a more complete video editing software if you'd like to slow it down even further and examine it better.

Could it be an advanced civilization trying to send a message and destroying the rocket because the fuel we are using is far too inefficient for fast and long distance travel? Could it be because we're developing technology that could allow us to get closer to them and they don't want that? Or could it merely be something that conveniently got into the shot at just the right moment and is nothing out of the ordinary? Either way, it is definitely interesting and will make you wonder.

Let's assume there is other life in the universe. If they're less advanced than us, we will likely have to find them. If there is an advanced civilization elsewhere in the universe, the chances are that they're already observing us. If so, they would have to have come from a planet that is farther than we can see in any detail, far beyond our own solar system.

To travel from such a great distance, they would have to have far more advanced technology than us and would undoubtedly be super-intelligent. Being that much more advanced than us, they may be able to affect our reality with advanced science or higher dimensions. This would allow for the Mandela Effect to be theoretically possible, but I doubt an advanced civilization would have any interest in changing Chick-fil-A logos and movie lines from our history unless they have a very peculiar sense of humor.

Maybe they're observing us and waiting for the opportune moment to reveal themselves. I highly doubt that they will reveal themselves and help us by offering technology, information and resources that would sling-shot us centuries into the future, at least not until we can all get along and learn to work together.

If they are up there, we probably need them a lot more than they need us, so maybe we should strive to contribute to the success and survival of our world and a sustainable future so we can offer some sort of value and reason for them to want to disclose their existence to us.

Are We A Simulation?

Another question that has come up a lot lately is: "Are we a simulation?" or "Are we living in a simulated reality?".

There is no way to prove or disprove whether we are part of a simulation, but billionaire entrepreneur and founder of SpaceX, Elon Musk, shared his thoughts on this subject that covers the entire theory until we can prove or disprove it. I will not paraphrase what he said, but I will give you the gist...

40 years ago, our top-of-the-line video games were things like "*Pong*" (two squares and a circle). Now, just a few decades later, which is nothing in the timeline of humanity, we have near photo-realistic video games, with millions of people playing simultaneously online, from all over the world. Virtual reality is gaining traction now. With our progress in gaming, we will soon create virtual reality (a simulated world) that will be indistinguishable from our own and we could create sentient characters that would live in a VR world without knowing it.

Elon then asked: *"How do we know this hasn't happened before?"*

A glitch in the Matrix would be easy to believe if this happened to be the case with us. Our creators/designers would have fabricated our history, as well as having given us what we know as "consciousness". A change in our history would consist of a few simple clicks of a button and the changing of a piece of code, just like we have the ability to do with our own video games, and we'd never really know the difference.

Unfortunately, if this is true, the only way we could break out of a virtual reality or simulated world, would be to become more advanced than the civilization that created us. I can't imagine that is very likely.

The Time Traveling Girl

There is a pretty convincing story about a 5-year-old girl who was on a family adventure and at some point, walked into a certain part of Hoia Baciu Forest, just outside of Cluj-Napoca, in North West, Transylvania, Romania, and then completely disappeared. Extensive searches were carried out, but she was gone.

Five years later, locals found a 5-year-old girl on the outskirts of the forest. It was the girl that had gone missing half a decade before. She was wearing the same clothing she'd had on the day she went missing and had not aged at all.

This is not the only case where people have reported lost time in the area. Though there isn't a whole lot of information on these reports, if true, it's possible that it could be a portal to another world, dimension or something even more bizarre. The forest is also a hot bed for UFO and paranormal activity. With the possibility of these portals existing in our reality, we may not have even scratched the surface of understanding the universe.

FINAL THOUGHTS

The reasons I decided to write a sequel were due to the popularity around the Mandela Effect, but mainly to offer a logical and down-to-earth conclusion to the phenomenon for people that were still searching for answers.

Many people, including myself, have lost sleep over this phenomenon. Raking through every quadrant of our brains to find answers and recall the actual events of what we *thought* we remembered. Some have even suffered to such a degree from the Mandela Effect phenomenon that it caused them to feel depressed and even suicidal. There are people who have reported that they have crossed into another reality and the life they have always known no longer exists. Some have even had to resort to prescription medications to deal with the emotional and psychological effects that they've experienced from it.

For some of us, the Mandela Effect is just a

cool party trick to blow the minds of your friends. For others, it has become a fascination that has led them into a deep investigation of theories of parallel universes and alternate dimensions. But for some, it has caused real, psychological trauma.

I wanted to share my conclusions to this worldwide phenomenon to help people understand that they are not going insane, they are not in someone else's body, and there are logical answers and reasons to why many of us experience these strange, alternate memories.

I sincerely hope I have provided you with some conclusive answers and explanations to those affected by the phenomenon and especially to those experiencing genuine psychological distress with the Mandela Effect.

*Just because it's unexplained,
doesn't mean it can't be explained.*

ANSWERS TO LOGICAL REASONING

1. Half way, because then it would be running out of the woods.
2. A green house is made from glass, not bricks.
3. There is no soil in a hole.
4. He stood on a block of ice and the ice melted.
5. Whatever color *your* eyes are. The question start with "YOU fly a plane".
6. There is no such thing as a melon tree.
7. Moses didn't take any animals on the ark, Noah did.
8. There are no stairs in a single-story home.

Available from Amazon, Barnes & Noble, CreateSpace & StockRoomDeals

www.ingramcontent.com/pod-product-compliance
Lightning Source LLC
Chambersburg PA
CBHW070108210526
45170CB00013B/791